# HORSE FARMING FOR BEGINNERS

Essential Tips On Breeding, Care, And Training For Successful Equine Management

**Holden Bodhi**

# Contents

**CHAPTER ONE** ................................................................ 9
  Overview Of Equine Agriculture .............................................. 9
    An Overview Of The Advantages Of Horse Farming .......... 9
    Recognizing The Various Kinds Of Horse Farms ............. 11
    Principles Of Equine Health And Welfare ...................... 14

**CHAPTER TWO** ............................................................. 17
  Selecting The Appropriate Breed ......................................... 17
    Things To Take Into Account While Choosing A Breed .... 17
    Popular Breeds To Introduce To Beginners ..................... 20
    Breed Selection Based On Purpose .................................. 22

**CHAPTER THREE** ......................................................... 25
  Establishing A Farm ............................................................ 25
    Crucial Infrastructure ...................................................... 25
    Making Safety And Accessibility Plans ............................ 29
    Basic Tools And Equipment Required ............................. 31

**CHAPTER FOUR** .......................................................... 35
  Nutrition And Feeding ........................................................ 35
    Knowing Your Horse's Nutritional Needs ....................... 35
    Comprehending Nutritional Needs .................................. 36
    Feeding Plans And Optimal Techniques ......................... 39

**CHAPTER FIVE** ............................................................ 43
  Medical And Veterinary Services ....................................... 43
    Standardized Procedures In Healthcare ......................... 43
    Typical Health Problems And Their Avoidance .............. 46

Building A Connection With A Veterinarian .................... 48

**CHAPTER SIX** .................................................................. 51

   Cleaning And Organising ..................................................... 51

      Essential Grooming Methods And Equipment ............. 51

      Taking Care Of And Education Young Horses ................ 54

      Developing Communication And Trust With Horses ....... 55

**CHAPTER SEVEN** ............................................................. 59

   Basic Training ..................................................................... 59

      Recognizing The Psychology And Behaviour Of Horses ... 59

      Outlining The Foundation And Fundamental Orders ...... 62

      Creating Practical Training Objectives ........................... 65

**CHAPTER EIGHT** ............................................................. 69

   Breeding Factors To Consider In Equine Farming ................ 69

      The Fundamentals Of Horse Breeding ........................... 69

      Ovulation And Timing ................................................... 72

**CHAPTER NINE** .............................................................. 78

   Regulation And Law Requirements ..................................... 78

      Recognising Horse Farming Zoning Laws ....................... 78

      Licenses And Permits Required ...................................... 80

      Insurance Things To Think About For Horse Farms ......... 81

**CHAPTER TEN** ................................................................ 86

   Promoting Your Equine Farm .............................................. 86

      Deciding Who Your Target Market Is .............................. 86

      Marketing Your Services (Training, Breeding, And Boarding) ....................................................................... 88

Creating An Easy-To-Use Website ................................... 91

Making use of social media ............................................ 91

CHAPTER ELEVEN .................................................................. 94

Increasing Your Expertise And Capabilities ......................... 94

Sources For Additional Education .................................... 94

Developing Professional And Horse Farming Networks .. 96

Keeping Up With Best Practices And Industry Trends ..... 98

Copyright © 2024 by Holden bodhi

All rights reserved.

No part of this publication may be reproduced, distributed, or transmitted in any form or by any means, including photocopying, recording, or other electronic or mechanical methods, without the prior written permission of the publisher, except in the case of brief quotations embodied in critical reviews and certain other noncommercial uses permitted by copyright law.

## DISCLAIMER

The information provided in this book, is intended for educational and informational purposes only. The content is based on research, personal experiences, and general knowledge about farming. It is not intended to substitute professional advice or expert consultation. Readers are encouraged to seek professional guidance when implementing any practices or techniques discussed in this book.

The author and publisher make no representations or warranties of any kind regarding the accuracy, applicability, or completeness of the contents of this book. Any reliance you place on such information is strictly at your own risk. The author and publisher shall not be held liable for any damages, losses, or injuries resulting from the use of the information provided.

Additionally, the author does not endorse, recommend, or affiliate with any individual, product, service, website, organization, or brand mentioned or referenced in this book. Any such references are solely for informational purposes, and no warranty or guarantee is implied. The inclusion of these references does not imply any endorsement or partnership by the author.

By reading this book, you acknowledge and accept that the author and publisher are not responsible for any consequences arising from your use of the information provided

# CHAPTER ONE

## Overview Of Equine Agriculture

Combining the excitement of equine companionship with the benefits of agriculture is horse farming, an exciting and fulfilling endeavor. Over the ages, this age-old custom has undergone tremendous change and grown to be an essential component of numerous civilizations across the globe. Entering the realm of horse husbandry provides a rare chance to bond with these amazing creatures and gain many rewards for those who do so.

## An Overview Of The Advantages Of Horse Farming

Horse breeding, raising, and management are the main components of horse husbandry. Aside from the obvious happiness that comes with raising these creatures, horse farming has several advantages for the economy, society, and individuals. The possibility of making money is one of the main benefits. With the popularity of horseback riding, equestrian sports, and leisure riding growing a well-run horse farm can make a sizable profit from selling horses, providing training services, giving riding lessons, and even selling boarding facilities.

Furthermore, improving land management with horse farming may be a successful strategy. Through grazing, horses can contribute to the upkeep of pastures, ensuring their sustainability and health. By reducing the need for chemical pesticides and fertilizers, this technique encourages the utilization of land in an environmentally sustainable manner. Horses are also renowned for their capacity to build enduring bonds with people. Interacting with horses can improve mental health by offering companionship and emotional support. Spending time with their horses has been shown to reduce stress and boost happiness in many horse owners.

Horse farming can also promote social cohesion and community. Through participation in local riding groups, events, or farm visits, horse owners frequently become members of a wider community of equestrian enthusiasts. In order to make horse farming a rewarding endeavor on both a personal and professional level, this group may provide priceless support, knowledge sharing, and friendship.

## Recognizing The Various Kinds Of Horse Farms

Horse farms are available in a variety of sizes and forms, each meeting unique requirements, tastes, and objectives. Anyone wishing to begin their career in horse husbandry must have a thorough understanding of these many kinds.

1. Farms that specialize in breeding horses: These farms strive to produce high-quality

horses, usually in a specific breed. An in-depth understanding of genetics, breeding procedures, and how to care for expectant mares and their foals are all necessary on breeding farms.

2. Training Facilities: Dressage, jumping, and racing are just a few of the disciplines for which training farms prepare their horses. To guarantee that the horses receive the best training possible, these farms frequently hire seasoned riders and trainers.

3. Boarding Stables: For equestrians who might lack the space to keep their horses at home, boarding farms offer services. These farms usually provide tack, feed, and care services, so horse owners may relax knowing their animals are in capable hands.

4. Riding Schools: Beginners and aficionados wishing to take up horseback riding can find instruction at these farms. All ages and skill

levels are often catered for in riding schools, which also typically offer a variety of school horses that are well-mannered and secure for inexperienced riders.

5. Rescue and Rehabilitation Facilities: A few equestrian ranches concentrate on saving and healing underprivileged horses. These facilities are essential for giving neglected or abused horses a safe sanctuary, assisting in their health restoration, and helping locate their new homes.

6. Eventing & Competition Farms: These farms cater to riders who want to compete, and they frequently offer facilities and specialized training to get riders and horses ready for shows. They concentrate on several different disciplines, such as show jumping, cross-country, and dressage.

## Principles Of Equine Health And Welfare

Horse care is a complex undertaking that needs commitment, expertise, and the right tools. Any horse farmer who wants to ensure the health and welfare of the animals in their care must have a basic understanding of horse care and management.

1. Nutrition and Feeding: For optimal health, horses need a well-balanced diet. Hay, grains, and supplements based on their age, weight, and activity level usually make up this. Anyone interested in horse husbandry must have a basic understanding of the nutritional requirements of horses and how to address those needs.

2. Health Care: Keeping a horse healthy requires routine veterinarian care. This covers shots, deworming, dental work, and routine examinations. For early intervention and

treatment, it is essential to recognize symptoms of disease or discomfort.

3. Regular grooming not only keeps a horse in good appearance but also aids in the detection of any possible health problems, such as injuries or skin disorders. To keep a horse healthy and sound, proper foot care is also essential, which includes routine trimming and shoeing.

4. Housing & Shelter: It's essential to provide horses with suitable shelter to keep them safe from inclement weather. Horses should have a safe, hygienic, and comfortable area to rest and hide from the weather in stalls or run-in shelters.

5. Exercise and Training: To maintain physical fitness and cerebral stimulation in horses, regular exercise and training are essential. To build a well-rounded horse, one must

comprehend the fundamentals of horse training, including groundwork and riding practices.

6. Horses are gregarious creatures who flourish in the companionship of other equines. Giving them socialization opportunities can improve their mental health, lower stress levels, and encourage a happy, healthy existence.

# CHAPTER TWO

## Selecting The Appropriate Breed

A novice horse farmer must make a critical choice when selecting their breed of horse. Your experience farming can be greatly impacted by the breed you choose because different breeds have distinct temperaments, physical attributes, and training requirements. In this section, we will examine the crucial elements to take into account when choosing a breed, present a few well-liked breeds that are appropriate for novices, and go over how to choose a breed depending on your objectives.

## Things To Take Into Account While Choosing A Breed

### 1. Level of Experience

Your degree of experience should be one of the first things you take into account when choosing a horse breed. Those who are new to the breed may prefer to choose those recognized for their easy training and serene temperament. New owners can develop their abilities and confidence with horses that are known to be beginner-friendly without having to deal with the extra strain of managing a more temperamental breed.

## 2. Ownership's Objective

Selecting the appropriate breed of horse requires knowing why you want a horse in the first place. Are you primarily seeking for a horse for work-related responsibilities, competitive sports, or pleasure riding? Different breed traits may be needed for each of these objectives. A mild trail horse, for example, would be very different from an extremely energetic competitive jumper.

## 3. Dimensions and Mass

Horses vary in size, so it's crucial to choose one that matches your physical skills. Smaller breeds could be easier for novices to handle, while larger breeds might need more room and more experienced hands. When selecting a breed, take into account your strength and size to make sure you can handle and take care of the horse.

## 4. Characteristics

One important factor to consider is the horse's disposition. While some breeds have a reputation for being quiet and gentle, others could be more gregarious and tense. Selecting a breed that is recognized for friendliness and ease might be beneficial for beginners as it can make learning less stressful and more pleasurable.

## 5. Needs for Upkeep and Care

Breed-specific care needs include those related to nutrition, grooming, and health. Certain breeds are more resilient and can survive with less intensive care, but others might need more regular trips to the vet or special diets. To make sure you can meet the criteria of the breeds you are considering, be careful to research their particular demands.

## Popular Breeds To Introduce To Beginners

### 1. Quarter Horse

The adaptability and amiable disposition of the American Quarter Horse make it one of the most favored breeds among novices. Because of their power and agility, they are well-suited for a variety of activities, including working cattle, trail riding, and competitive competitions. Their composed manner makes them an excellent option for inexperienced riders.

### 2. Horse Paint

Paint horses are renowned for their laid-back disposition and unusual coat patterns. They are bright, sociable, and adaptable, which makes them appropriate for a range of activities, including rodeo competitions and recreational riding. They are an ideal option for **novices** due to their gregarious disposition and eagerness to learn.

### 3. Morgan Horse

Morgan horses are renowned for their endurance, adaptability, and gentle disposition. They are perfect for driving, trail riding, and competitive activities. They are a good fit for inexperienced riders searching for a horse that will grow with them because of their enthusiasm to work and learn.

### 4. Haflinger

Small in stature but robust in build, the Haflinger is renowned for its kind nature. They are ideal for novices, especially kids, as they are simple to handle and teach. Haflingers are excellent in many areas, such as driving, riding, and even some competitive sports.

## 5. Arabian Horse

Although Arabian horses are commonly linked with endurance riding, they can also serve as fantastic partners for novice riders. They're a pleasure to work with because of their intelligence and loving disposition. They are better suited for novices who are eager to learn, nevertheless, as they could require more handling experience.

## Breed Selection Based On Purpose
### 1. Cycling

If recreational riding is your main objective, concentrate on calm, dependable breeds. Haflinger and Quarter Horse breeds are excellent choices for trail riding since they provide a comfortable ride and simple handling. When picking your choice, take into account things like your riding style and the riding places you intend to frequent.

## 2. Work

Because of their power and adaptability, breeds like the Quarter Horse and Clydesdale are perfect for anyone who wants to use horses for labor-intensive jobs like herding cattle or performing farm chores. Make careful to take the horse's temperament into account; a steady, placid horse will be more dependable in job settings.

## 3. rivalry

Some of the greatest breeds to consider if you want to participate in competitive activities or riding are those recognized for their performance and agility, such as the Arabian or Thoroughbred. These breeds frequently do well in a variety of sports, such as dressage and show jumping. Make sure the breed you choose fits your experience level and competitive objectives.

# CHAPTER THREE

## Establishing A Farm

Establishing a horse farm necessitates meticulous preparation and arrangement. Finding the ideal site and creating a room that satisfies your horses' needs while guaranteeing efficiency, comfort, and safety are the first steps in the process. This section describes the necessary infrastructure, accessibility, safety planning, and basic tools and equipment required to establish a profitable horse farm.

## Crucial Infrastructure

**Consistency**

The foundation of each horse farm is the stable. When your horses aren't in the pasture, they offer a secure and pleasant atmosphere. When planning your stables, keep the following in mind:

1. Size & Space: To provide enough room for the horse to walk about comfortably, each stall should measure at least 12 feet by 12 feet. Even more room may be needed for larger breeds. Furthermore, make sure that stables are at least 10 feet tall to allow for the natural stance of the horse.

2. Ventilation: Keeping the air clean and avoiding respiratory problems depends on proper ventilation. To encourage ventilation, include windows, vents, and fans in the design of your stable. The stable should have cross-ventilation to be dry in the winter and cool in the summer.

3. Bedding: For comfort and hygienic reasons, select suitable bedding materials such as wood shavings, rubber matting, or straw. To reduce smells, bedding should be absorbent and simple to clean.

4. Water Supply: To guarantee that your horses always have access to clean water, place water buckets or automated waterers in each stall.

**Pastures**

Your horses' well-being depends on their pastures, which give them an area to graze, exercise, and interact with one another. When creating pastures, consider the following:

1. Size and Layout: To avoid overgrazing and guarantee that each horse has ample room to roam, provide a minimum of one acre of pasture for each horse. Take into account grazing systems that rotate to help pastures heal.

2. Fencing: To keep your horses safe, choose strong, safe fencing materials like wood, vinyl, or electric fencing. To stop escapes, the height should be at least four to five feet.

3. Grass Varieties: Choose the right kinds of grass for the climate and region you live in.

Horses should be able to eat healthy, tasty grass, and it might need to be maintained on a regular basis with mowing and reseeding.

4. Shelters: To shield horses from severe weather, provide cover within the grazing area, such as run-in shelters.

**Arenas**

Arenas, also known as training areas, are necessary for riding and exercising your horses. When creating an arena, take into account the following factors:

1. Surface Material: Sand, rubber, or a combination that reduces the chance of injury while offering a good grip are suitable options for the surface, depending on the sport.

2. Dimensions: A conventional dressage riding arena measures 60 feet by 120 feet, whereas a jumping arena measures 100 feet by 200 feet.

Make sure there is enough room for all the training activities.

3. lights: Make sure you have adequate lights installed for use in the evening and on cloudy days. While natural light is ideal, electric lights could be necessary in addition.

4. Drainage: To keep water from collecting and to provide a safe surface, proper drainage is crucial. Provide suitable drainage systems and a modest slope in the arena's design.

## Making Safety And Accessibility Plans
### Safety Procedures

When setting up a horse farm, safety needs to be your first concern. Put the following actions into action:

1. Emergency Exits: Make sure the emergency exits in your arena, pastures, and stable are properly marked. Provide staff and family

members with regular emergency procedure training.

2. First Aid Kits: Provide basic medical supplies such as bandages, antiseptics, and pharmaceuticals in first aid kits. Check the kits frequently and replenish them as necessary.

3. Fire Safety: Set up extinguishers all over the building and practice fire drills on a regular basis. Create an evacuation plan as part of your fire safety strategy.

4. Frequent Maintenance: Examine all buildings, machinery, and fencing on a frequent basis. To keep things safe, take quick action to resolve any problems.

## Availability

Make sure that everyone, including people with impairments, may access your horse farm. Think about the following:

1. Walkways and Paths: Create broad, level walkways that make it simple to reach arenas, pastures, and stables. To maintain safety, use non-slip materials.

2. Parking: Give cars and trailers plenty of room to park close to the entrance. Larger trucks and trailers should be taken into account, particularly during events.

3. Restrooms: Provide personnel, guests, and visitors with accessible restrooms. Make sure they are close to the major amenities, spotless, and well-stocked.

## Basic Tools And Equipment Required
**Crucial Instruments**

A variety of items are needed to set up a horse farm, maintain the infrastructure, and take care of the horses. The following is a list of essential gear:

1. Grooming Supplies: Sponge, combs, brushes, and hoof picks are among the basic grooming supplies. Regular grooming is essential for the health and cleanliness of horses.

2. Feeding Equipment: Make a good investment in water troughs, hay nets, and feed buckets. Regular feeding schedules may also benefit from automatic feeders.

3. Cleaning Supplies: Pitchforks, shovels, rakes, and muck buckets are some of the items you'll need to keep your pastures and stables clean. Frequent cleaning keeps garbage from piling up and fosters a healthy atmosphere.

4. Transport Equipment: Whether taking horses to shows, doctor appointments, or pasture rotations, a horse trailer is needed. Make sure it's safe to travel in and well-maintained.

## Upkeep Tools

Ensuring the durability and safety of equipment and facilities requires routine maintenance. Typical maintenance instruments consist of:

1. Fencing Tools: Get post drivers, wire cutters, and hammers, among other tools, to help you install and maintain fencing.

2. Tools for landscaping: Use equipment like lawnmowers, weed eaters, and pruning shears to keep pastures and their surroundings in good condition.

3. Tools for Repairs: Keep an organized toolbox with a basic toolkit that includes pliers, screwdrivers, and wrenches.

## Particularised Tools

Depending on your objectives and areas of expertise, you might require specialized gear like:

1. Equipping for Riding: Helmets, bridles, and saddles are necessary for any kind of riding. Invest in high-quality equipment for comfort and safety.

2. Training Aids: Lunge lines, cones, and poles for different exercises are things to think about buying if you intend to teach horses.

3. Medical Supplies: For your horse's health, you'll need a fully equipped medical kit as well as any prescription drugs your veterinarian may recommend.

# CHAPTER FOUR

## Nutrition And Feeding

### Knowing Your Horse's Nutritional Needs

As herbivorous creatures, horses have adapted to spend the majority of the day grazing. Since their digestive systems are specifically made to break down fibrous plant materials, any horse farmer must understand the type of diet they should be feeding their horses. Horses, in contrast to many other animals, have tiny stomachs about their body size and a lengthy digestive tract, which means they require many feedings throughout the day.

**The Function of Fibrek**

A horse's diet must include fiber, which is mostly obtained from grass, hay, and other fodder. It facilitates healthy digestion and guards against ailments like laminitis and colic. Forage must provide at least 1% of a horse's body weight each day, or 15 to 20 pounds of hay for a typical-sized horse. The fiber should come from premium grass or legume hays for maximum health.

**Water Consumption**

Another vital component of a horse's diet is water. Horses need five to ten gallons of clean, fresh water each day, depending on their size, activity level, and surroundings. They should always have access to this water. Maintaining proper hydration is crucial since dehydration can cause major health problems, such as colic.

## Comprehending Nutritional Needs

Horses need a well-balanced diet that includes fats, carbohydrates, minerals, vitamins, and fiber. Grain is the main source of carbohydrates, which are necessary for energy, particularly in active horses. Although they are not as frequently included, fats can offer concentrated energy and are best obtained from supplements or vegetable oils.

Minerals and vitamins are essential for immune system support, general well-being, and health maintenance. To make sure they get enough of essential nutrients, horses frequently require supplements, particularly if their feed is inadequate.

## Feed and Supplement Types

Horse owners can supply their horses with balanced nutrition that is suited to each horse's unique needs by being aware of the many types of feed that are available and when to use them.

## Forage: The Basis of Diets for Horses

A horse's diet should consist primarily of forage. Grass and hay are included in this; hay varieties include clover, alfalfa, and timothy. Alfalfa is frequently higher in calcium and protein, making it advantageous for growing horses or lactating mares. Grass hay, on the other hand, has fewer calories and is perfect for helping less active horses maintain their weight.

## Pellets and Grains as Concentrates

Grain products with a high energy content, such as barley, corn, and oats, are known as concentrates. To provide energy needs, they are frequently combined with fodder, particularly for performance horses or horses with greater caloric needs. With a combination of grains, vitamins, and minerals, commercially prepared meals like pellets and complete feeds can make nutrition management easier.

## Supplements: Improving Nutrition

Even though most nutritional demands may be met by a well-balanced diet, several supplements can improve a horse's overall health and performance. Typical add-ons consist of:

- Mixtures of vitamins and minerals: These bridge dietary shortages in essential nutrients.

- Joint supplements: These can help performance horses' joints stay healthy since they contain glucosamine or chondroitin.

- Probiotics: Good microorganisms that enhance nutrient absorption and maintain intestinal health.

Electrolytes: Vital for recuperation and hydration, particularly in equestrian competition horses.

## Feeding Plans And Optimal Techniques

Maintaining a horse's health and well-being requires setting up an appropriate feeding regimen and adhering to best practices.

### How Often You Feed

Since horses are grazers by nature, a feeding plan that reflects this behavior is ideal for them. Feeding horses two or three times a day is recommended, with an emphasis on consistently giving grass. Slow feeders, if available, can help horses consume tiny amounts of hay throughout the day, which will help them be less bored and avoid stomach problems.

## Control of Portion

It's essential to keep an eye on portion sizes to avoid obesity and its related health issues. Feed amounts may need to be adjusted based on the size, age, and activity level of the horse. The horse's physical condition score can be used to guide diet modifications and portion amounts regularly.

## hygienic and secure feeding procedures

It's critical to keep feeding places clean. To avoid the formation of mold and germs, which can cause health problems, make sure that feed buckets and troughs are cleaned regularly. To avoid contamination and spoiling, feed should also be stored correctly in a cool, dry environment.

**Particular Points to Remember**

Certain horses, depending on their age, condition, or degree of exercise, could require a different diet. Senior horses may benefit from softer diets that are easier to chew and digest, but young, growing horses need higher protein levels for development. Horses with certain medical conditions, such as metabolic syndrome or laminitis, may need particular diets and need to be under a veterinarian's strict supervision.

**In summary**

Any newcomer to horse farming must have a thorough understanding of the nutritional and feeding requirements of horses. Horse longevity can be enhanced by feeding a well-balanced diet high in forage, choosing concentrates and supplements wisely, and adhering to a regular feeding schedule. Using best practices guarantees a nutritious and healthy diet, enabling your horses to flourish in their surroundings.

# CHAPTER FIVE

## Medical And Veterinary Services

Ensuring the happy and healthy lives of your horses depends on proper veterinarian treatment and health maintenance. The health of your horses will depend on your ability to grasp normal medical care, common health problems, and how to build a rapport with a veterinarian as a novice horse farmer.

## Standardized Procedures In Healthcare

Maintaining the general health and well-being of your horse depends on routine medical care. As a horse owner, you ought to arrange routine check-ups for several important medical procedures:

1. Schedule of Vaccinations: Horses should receive vaccinations as part of their preventive healthcare. Vaccines guard against several several dangerous illnesses, such as the West

Nile virus, tetanus, and equine influenza. To create a vaccine regimen specific to your horse's age, health, and the diseases that are common in your area, speak with your veterinarian.

2. Parasite control: Preventing parasite infestations requires routine deworming. Horses may carry a variety of internal and external parasites that are harmful to their well-being. Together with your veterinarian, come up with a deworming strategy that takes into account your horse's unique requirements and fecal egg counts.

3. Dental Care: To keep their teeth in good health and the right alignment, horses need to have regular dental examinations. Dental problems can cause discomfort and make eating difficult, which might affect their general health. Routine dental examinations should be carried out by a licensed veterinarian or equine dentist at least once a year.

4. foot Care: The comfort and mobility of your horse depend on proper foot care. Every 6 to 8 weeks, a qualified farrier should do routine hoof trimming and shoeing (if required). Check the hooves of your horse regularly regularly for cracks, thrush, or other problems.

5. Feeding and nutrition: The cornerstone of your horse's health is a well-balanced diet. To design a feeding schedule that suits your horse's nutritional requirements according to age, weight, and activity level, speak with a veterinarian or equine nutritionist. Make sure your horse has access to pasture or high-quality feed in addition to clean, fresh water.

6. Exercise and Turnout: Staying in a healthy weight range and being physically fit requires regular exercise. Create a schedule for your horse that incorporates mental and physical stimulation, based on breed and intended use. Horses benefit mentally from having time to socialize and graze during turnout.

## Typical Health Problems And Their Avoidance

Keeping your horse healthy can be achieved by taking preventive steps and being aware of common health conditions that affect horses. The following list of common health problems and preventative measures:

1. Abdominal pain, or "colic," can be caused by several factors, such as dietary modifications, dehydration, or intestinal blockage. Keep a regular feeding schedule, give your child unrestricted access to water, and refrain from making abrupt dietary changes to prevent colic.

2. Laminitis: Overindulging in grains or gaining too much weight are two common causes of this excruciating illness that affects the hooves. Keep an eye on your horse's diet, give his hooves frequent attention, and keep him at a healthy weight to avoid laminitis.

3. Problems with the Respiratory System: Allergens, inadequate airflow, and respiratory infections all cause respiratory issues in horses. Keep the barn tidy, make sure there is adequate ventilation, and stay away from dusty feeds to reduce these risks.

4. Skin Conditions: Irritating substances or inadequate cleanliness can lead to skin conditions like dermatitis and fungal infections. These ailments can be avoided with good hygiene, appropriate wound care, and regular grooming.

5. Cushing's Disease: A common condition in elderly horses, Cushing's disease disrupts hormone balance and can result in several health problems. Frequent veterinary examinations can aid in identifying the condition's early symptoms and starting treatment plans.

## Building A Connection With A Veterinarian

As a horse owner, one of the best investments you can make is a good working relationship with your veterinarian. The following advice can help you form and keep a fruitful partnership:

1. Select the Correct Veterinarian: Seek out a medical professional who has knowledge of veterinary treatment for horses and who is sensitive to your unique needs as a horse owner. Consult with nearby equestrian organizations or other horse owners for recommendations.

2. Plan Frequent Check-Ups: Early diagnosis of health issues is facilitated by routine veterinary appointments. To keep your horse healthy, make it a priority to arrange yearly check-ups, vaccines, and dental exams.

3. Keep Lines of Communication Open: Establish a rapport with your veterinarian by

talking about the health, manners, and any worries you may have regarding your horse. Giving them precise information will enable them to customize care to your horse's individual needs.

4. Emergency Preparedness: Keep your veterinarian's contact information close to hand in case of an emergency. Talk about emergency procedures and what to do in the event of a medical emergency.

5. Take Advice to Heart: Pay attention to what your veterinarian has to say and be willing to learn. They may offer insightful advice on diet, exercise, and general equine care that will improve the health of your horse.

6. Act Proactively: Get in touch with your veterinarian as soon as problems emerge. Keep a regular eye on your horse's behavior, eating patterns, and general health. Consult your

veterinarian for advice if you observe any changes.

You can make sure that your horses remain healthy, content, and prepared for their tasks on your farm by concentrating on these important aspects of health and veterinary care. A healthy horse farming environment can be established with the assistance of regular health care procedures, awareness of frequent health problems, and a solid rapport with your veterinarian.

# CHAPTER SIX

## Cleaning And Organising

The health and welfare of your horses are dependent on proper grooming and handling, two essential components of horse farming. These behaviors not only keep horses in peak condition but also strengthen the relationship between the handler and the horse. This part will cover basic grooming methods and supplies, the value of handling and training young horses, and tactics for fostering communication and trust.

## Essential Grooming Methods And Equipment

Beyond appearance, grooming is a vital component of horse care. Frequent grooming offers the chance to check for injuries or abnormalities, avoid health problems, and maintain the health of the horse's skin and coat.

1. Brushing Techniques: To dislodge filth, muck, and loose hair, start with a curry comb, or a rubber or plastic instrument. To stimulate the skin and encourage blood circulation, move in circular patterns. To get rid of any remaining dirt and debris, use a stiff-bristled brush after using the curry comb. A soft brush is the best choice for delicate areas like the face and legs to minimize irritation.

2. Hoof Care: Keeping hooves clean and in good condition is essential to avoiding hoof-related problems. After clearing the hoof of dirt and debris using a hoof pick, look for any stones or other foreign items. Taking good care of your horse's hooves can help prevent diseases like laminitis and thrush, keeping them sound and healthy.

3. Bathing: Although horses don't need to be bathed frequently, giving them an occasional

wash will assist get rid of perspiration and grime. To avoid irritation, use a gentle shampoo made especially for horses, and be sure to rinse well. To prevent chills, dry your horse off thoroughly after bathing, especially in colder months.

4. Mane and Tail Maintenance: To avoid breakage, tangles in the mane and tail must be worked out. To prevent pulling and irritation, use a wide-toothed comb and work gently from the bottom upwards. To make things easier and keep the hair in control, think about applying a detangling spray.

5. Examining your horse's health during a grooming session is a great way to see whether there are any wounds, edema, or skin issues. Observe the state of the coat, the eyes' clarity, and any strange behaviors that might point to a disease or discomfort.

## Taking Care Of And Education Young Horses

The care and education of young horses are essential to their growth and future abilities. To safely and efficiently guide young horses, one needs to possess patience, consistency, and knowledge.

1. First Handling: Begin by acquainting young horses with lead ropes and halters. Teach them to react to pressure gradually; this is a necessary skill for safe handling. Treats and praise are examples of positive reinforcement tactics that can be used to assist in establishing a positive relationship with handling.

2. Desensitisation: New situations and experiences can make horses sensitive. To help young horses become less afraid and more adaptive, expose them to a variety of stimuli, such as sounds, objects, and other animals. It's important to expose them

gradually; don't expose them to too much at once.

3. Ground Training: A well-mannered horse starts with ground training. Teach simple orders with consistent verbal cues and body language, such as "whoa," "walk on," and "back." You and the horse develop mutual respect and understanding as a result of this training.

4. Training Longevity: To avoid dissatisfaction and exhaustion, concentrate on brief, constructive training sessions. Engaged, not overwhelmed, is the optimum learning environment for horses. To keep children active, strike a mix between work and play by including lunges, foundation, and other activities.

## Developing Communication And Trust With Horses

Establishing a successful partnership with your horse requires regular communication and trust-building. For both the horse and the handler, a trustworthy relationship results in improved performance and a more pleasurable experience.

1. Recognising Body Language in Horses: Body language is the main way that horses communicate. Recognize how to read their cues, like where their ears are, how their tails move, and how they look. To handle and manage your horse effectively, you must be aware of his or her comfort level and mood.

2. Routine and constancy: Horses are creatures of habit and constancy. Create a routine for training, grooming, and feeding your horse so that he feels safe and knows what's expected of him. You may develop trust and reinforce learning by being consistent in your commands and behaviors.

3. Positive Reinforcement: Reward desired behaviors using verbal praise or treats, among

other techniques. By using this method, you can get your horse to repeat the behavior and fortify your relationship.

4. Empathy and patience are necessary since developing a trustworthy bond with your horse takes time. Have patience with them and show empathy for their feelings. In order to help them feel comfortable and secure around you, acknowledge their anxieties and gradually desensitize them.

5. Establishing a Positive Environment: Make sure your horse's surroundings are secure and cozy. A calm environment eases tension and builds trust, so your horse feels safe and comfortable with you.

2.         .

# CHAPTER SEVEN

## Basic Training

### Recognizing The Psychology And Behaviour Of Horses

Horses are intelligent animals with distinctive behavioral characteristics influenced by their social systems, biology, and innate tendencies. For management and training to be effective, it is imperative to comprehend these elements.

### 1. innate instincts

Horses have evolved to be alert and responsive to their environment since they are predatory animals. They react strongly to changes in their surroundings because of their fight-or-flight reaction. Acknowledging these inclinations promotes trust and reduces anxiety when training.

## 2. Animals with Social Behaviour

Because they are such gregarious creatures, horses frequently do well in herd settings. They create social hierarchies and use vocalizations, body language, and even nuanced facial expressions to communicate. Gaining an understanding of these social factors can improve the way you communicate with your horse and create a pleasant training environment.

## 3. Styles of Learning

Horses pick up new skills via repetition and experience. Both positive and negative reinforcement elicit a response from them. While negative reinforcement—like pressure—can aid in imparting orders, positive reinforcement—like treats or praise—encourages favorable behaviors. You may adapt your training techniques to your horse's demands by understanding their preferred learning style.

## 4. Body Language

It is essential to comprehend equine body language to communicate effectively. Horses communicate their feelings through their stance, eyes, and ears. For instance, a calm horse will have its ears forward, but an angry or nervous horse may have its ears pinched back. Being aware of these cues strengthens relationships and helps avoid misunderstandings.

## 5. Establishing Trust

Building trust with your horse is a process that takes time and requires respect, consistency, and patience. To establish a rapport, spend time with your horse outside of training sessions. Grooming, handling, and just being there are all included in this. An effective training relationship is built on trust.

## Outlining The Foundation And Fundamental Orders

In order to prepare a horse for future riding and handling, groundwork is a crucial part of horse training. Without putting too much weight on the rider, it focuses on building trust and communication between the horse and the handler.

### 1. The Value of Preparation

Establishing mutual respect and understanding with your horse via groundwork is beneficial. It gives you a controlled environment to evaluate the horse's temperament and behavior while encouraging it to respond to your suggestions. By doing this, you can increase your confidence and lower your risk of riding-related incidents.

### 2. Fundamental Orders

To create control and communication, start with basic commands. Among the important commands are:

- Stop: To get the horse to stop, use a firm voice and a light tug on the lead rope. Give them something for their cooperation.

- Move ahead: Encourage the horse to move ahead by lightly pressing on the lead rope or halter. Apply more pressure gradually until the horse reacts.

- Back Up: As you take a step back, stand facing the horse and softly tug on the lead rope. Your horse ought to follow you by default. Apply gentle pressure to reaffirm this directive.

- Turn: Use the lead rope and body language to teach your horse to turn. As you gently urge them to turn in the direction you want, step to the side.

3. De-sensitization

Desensitization is another benefit of groundwork; it helps your horse get used to different stimuli. Introduce children to things, sounds, and motions they could come across in daily life. Their reactivity while riding is enhanced and anxiety is decreased by this process.

## 4. The Secret Is Consistency

Training requires consistency. Every time you work with your horse, give the same commands, use the same body language, and give rewards. This constancy gives your horse a sense of security and reinforces what they've learned.

## 5. Brief Meetings

To keep your horse interested in training, keep your sessions brief and interesting. Aim for 20 to 30 minutes of concentrated work, interspersed with pauses. This method makes learning fun by preventing frustration for both the handler and the horse.

## Creating Practical Training Objectives

To keep yourself and your horse motivated, it's critical to set reasonable training objectives and monitor your progress. It guarantees that your training is well-organized, targeted, and doable.

### 1. Determine Your Horse's Skill Level

Determine your horse's current training and experience level before establishing goals. Take into account their breed, age, temperament, and training background. Knowing where they are coming from enables you to make goals that are suitable and meet their specific requirements.

### 2. Establish Specific Goals

Divide your training objectives into specific, quantifiable targets. Clear targets give focus and make it easy to track progress. For example, instead of saying, "I want my horse to be better at groundwork," specify, "I want my horse to respond to the 'back up' command consistently within three weeks."

### 3. Set Prioritised Objectives

Set your priorities according to your personal experience and the demands of your horse. Begin by mastering fundamental abilities like leading and stopping, then move on to more difficult ones like riding or lateral motions. This methodical approach creates a solid foundation and avoids overwhelm.

### 4. Be Adaptable

Although having a strategy is crucial, you should also be adaptable and ready to change course when necessary. Either horses can have bad days, or you can face unforeseen difficulties. Acknowledging this adaptability makes training more enjoyable and encourages long-term development.

## 5. Honour accomplishments

Celebrate your victories, no matter how tiny, as you accomplish your goals. Acknowledging your success gives you and your horse more self-assurance, strengthening your relationship and inspiring you to exercise more.

You may design a training program that is both successful and pleasant, and that strengthens the link between you and your horse by learning about horse behavior, implementing groundwork, and creating realistic training goals. This will be the starting point for a successful horse husbandry endeavor.

# CHAPTER EIGHT

## Breeding Factors To Consider In Equine Farming

### The Fundamentals Of Horse Breeding

To create children with desired features, horse breeding is a complex procedure that involves the careful selection of sires, male horses, dams, or female horses. Enhancing the caliber of horses for different sports, such as dressage, show jumping, and racing, requires this procedure. Before diving into the specifics of breeding, it's critical to grasp certain basic principles.

**Trait Selection and Genetics**

The performance and general quality of a horse are significantly influenced by its genetic composition. Breeders need to think about the features they want to increase or decrease in their kids as well as the genetic potential of the parents. These characteristics may include conformation, disposition, athletic prowess, and particular abilities pertinent to the horse's intended use. It's critical to have a thorough understanding of the pedigree and track record of prospective sires and dams.

## Selecting Ideal Breeding Stock

In order to choose the best breeding stock, prospective sires and dams must be assessed according to their health, performance history, and lineage. Horses having a track record of success in competitions or other similar activities are highly sought after by breeders. Genetic tests and health screenings can reveal important information about possible inherited conditions that may have an impact on the performance and well-being of the progeny.

## Planning and Breeding Objectives

It takes defined aims to breed horses successfully. Breeders should set precise standards for the ideal progeny, regardless of the goal: a champion racehorse, a dependable trail horse, or a competitive dressage mount. This encompasses not only physical characteristics but also disposition and simplicity of training. These objectives, the standards for choosing breeding stock, and a schedule for the breeding procedure should all be included in a breeding plan.

## Knowing How to Breed

There are multiple crucial phases in the breeding process, and each one needs to be carefully managed and observed.

## Mating Methods

A variety of techniques, including as live cover, artificial insemination (AI), and embryo transfer, can be used to breed horses. The conventional approach, in which the mare and stallion mate spontaneously, is known as live cover. However, the ease of use and potential to increase genetic diversity through the use of frozen or cooled semen from distant stallions have made AI increasingly attractive.

## Ovulation And Timing

Comprehending the reproductive cycle of a mare is vital for efficacious breeding. Mares are polyestrous, meaning that throughout the mating season, usually in the spring and summer, they go into heat (oestrus) more than once. The window around ovulation is crucial because breeding should take place while the mare is nearing ovulation and in heat. Periodic veterinarian tests, like as ultrasounds, can be used to determine the best time to breed.

### gestation and pregnancy

The mare will enter the gestation phase, which usually lasts roughly 11 months after she has successfully given birth. It is critical to keep a close eye on her nutrition and general health at this period. A well-balanced meal high in vitamins and minerals is necessary for pregnant mares to assist the growing fetus. Frequent veterinary examinations can aid in the early detection of any issues.

## Pregnancy Care For Foals And Mares

The vitality and welfare of the expectant mare and her youngster are critical components of an effective breeding operation.

## Monitoring of Nutrition and Health

A healthy diet is essential for the mare to have during her pregnancy. The increased energy requirements of gestation can be partially met by a well-balanced diet that includes premium forage, grains, and supplements. In addition, routine vet visits are necessary to keep an eye on the mare's health, including parasite prevention and immunizations.

## Getting Ready for Foaling

The mare should start getting ready for foaling as her due date draws near. This involves preparing a hygienic and secure space for the mare to give birth. Foaling stalls ought to be roomy, have good ventilation, and be furnished with the essentials, such as fresh bedding and a foaling kit that contains towels, thermometers, and disinfectant.

## Following Failure Care

As soon as the foal is born, it must get vital care. In the first few hours of life, the foal should be able to stand and nurse. The first milk a mare produces, called colostrum, is full of antibodies and is vital to a foal's immune system. After foaling, it's critical to keep an eye on the mare and the foal to make sure they're both doing well and recovering from their ordeal.

## Training and Socialisation

Early socialization is critical to the foal's growth as it matures. Introducing the foal to a variety of stimuli, including distinct settings, sounds, and other animals, can aid in its development into a well-mannered horse. Early basic training is also recommended, with an emphasis on handling and leading to getting the foal ready for further training in particular disciplines.

In summary

Any aspiring horse farmer must comprehend the complexities of horse breeding. Horse owners may ensure a successful breeding program that produces healthy and capable horses by understanding the breeding process, paying close attention to breeding concerns, and giving pregnant mares and their foals exceptional care.

# CHAPTER NINE

## Regulation And Law Requirements

Although raising horses is a fulfilling activity, there are many legal and regulatory requirements. Establishing a profitable horse farm and maintaining compliance with local, state, and federal laws require an understanding of these standards.

## Recognising Horse Farming Zoning Laws

The legality of running a horse farm on a certain plot of land is mostly dependent on zoning regulations. These rules are intended to control the use of land and guarantee that farming operations, including horseback riding, live in harmony with both residential and commercial properties.

### Zoning laws in the area

Checking with your local zoning authority is the first step toward establishing a horse farm. The majority of towns have zoning regulations and maps that outline the specific areas set aside for agricultural usage. Important things to think about are as follows:

• Zoning Districts: Depending on the district, there can be particular rules governing the size of buildings, the quantity of animals permitted, and the distances between properties. For instance, horse farming may be permitted in an area designated for agricultural use but not in a residential zone.

• Conditional Use Permits: You may be able to run a horse farm in an area that is not primarily meant for agricultural use in certain circumstances by obtaining a conditional use permit. Typically, this calls for the submission of a thorough plan detailing your proposed actions and proving they won't have a detrimental effect on the neighborhood.

• Future Modifications: Over time, zoning laws may change. It's wise to keep up with any proposed changes that could have an impact on your farming activities.

## Licenses And Permits Required

After confirming that your land is allocated for horse farming, you'll need to apply for several licenses and permits. The particular criteria may differ substantially according to your area and the size of your business.

**Municipal and State Permits**

• Business License: For anyone operating a commercial enterprise, the majority of municipal governments demand a general business license. This license attests to your company's compliance with regional laws.

• Agricultural Use Permits: Depending on your state, engaging in agricultural activity may require a certain permit. With this permit, you may be guaranteed that your farming methods adhere to safety and environmental regulations.

• Building permissions: You will probably want building permissions if you intend to build barns, stables, or other structures on your farm. These permits guarantee that your buildings adhere to zoning laws and safety standards.

• Health Permits: You could need health permits if some of the activities on your horse farm involve producing food, such as breeding horses for human consumption or vending equine-related goods.

## Insurance Things To Think About For Horse Farms

For any horse farm, insurance is an essential part of risk management. Working with animals entails inherent hazards, so it's critical to have the appropriate coverage to safeguard your investment and reduce potential damages.

**Insurance Types**

1. Liability insurance: If there is an incident or harm involving your horses, you are covered by this plan. Liability insurance can assist in paying for medical expenses and attorney fees in the event that a guest is hurt while on your property.

2. Property Insurance: Property insurance protects your equipment and structures from harm. This can involve defense against robbery, vandalism, and natural calamities.

3. Equine insurance: Your horses' health and welfare are covered by this specific type of insurance. It may cover medical expenses, death insurance, and liability for harm your horses may do.

4. Workers' compensation insurance covers medical costs and missed earnings for employees hurt on the job. If you hire people, you could be obliged to have this insurance.

## Selecting the Appropriate Insurance

When choosing an insurance plan for your equestrian farm, it's critical to:

• Evaluate Your Risks: Recognise the particular hazards connected to your business activities and select insurance appropriately.

• Consult Experts: To be sure you have the right coverage, work with insurance brokers who specialize in horse or agricultural insurance.

• Regularly Review Your plans: Your insurance plans should be reviewed as your farm develops and changes. You can modify coverage to suit your changing needs with the aid of regular reviews.

In summary

At first, navigating the legal and regulatory environment of horse farming may appear difficult. However, you may lay a strong basis for your horse farming endeavor by being aware of zoning regulations, obtaining the required licenses and permissions, and obtaining the right insurance coverage. Long-term success and sustainability in your business operations can be achieved by making the effort to make sure you comply with these regulatory standards.

# CHAPTER TEN

## Promoting Your Equine Farm

Successful marketing is essential to your horse farm's success. Whether you provide breeding, boarding, or instruction, knowing how to connect with and engage your target market can have a big impact on your company. Choosing your target market, advertising your products, and creating a powerful web presence are all covered in this part.

## Deciding Who Your Target Market Is

Determining your target market is the first step in marketing your horse farm. You may better target your marketing efforts to your target audience's requirements and preferences by getting to know who your potential consumers are.

**Recognizing Demographics** Analyze your local area's demographics first. Take into account variables like age, income, and way of life. Understanding the local community might help you identify potential clients because horse ownership frequently draws people and families who have a strong interest in equestrian activities.

## Dividing Your Viewership

Once you understand demographics, divide your audience into various groups. If you teach riding lessons, for example, your audience might include parents looking to teach their kids to ride as well as experienced riders. However, if you offer boarding services, horse owners searching for a dependable facility to take care of their animals may be your target market. You can develop marketing tactics that are more narrowly targeted by segmenting your audience.

## Surveys and feedback are being conducted.

Finding out the precise interests and preferences of your target audience may be accomplished through the use of surveys. Enquire about the experiences and services that current clients find most valuable. To learn more about what prospective customers want from a horse farm, interact with them at regional horse shows or on social media. To improve your marketing plan and make sure you are fulfilling the needs of your audience, feedback is crucial.

## Marketing Your Services (Training, Breeding, And Boarding)

The next stage is to advertise the services your horse farm provides after you have determined who your target market is. For each service to successfully draw in the relevant clients, a customized strategy would be needed.

### Boarding Facilities

In marketing your boarding services, highlight the special features of your establishment. Emphasize the caliber of care given, the dimensions of the paddocks, and any amenities that the horses can use, like trail access, veterinary treatment, or grooming services. Utilise client endorsements to establish credibility and highlight the advantages of boarding at your farm.

## Riding Instructions

Marketing for riding lessons should highlight the credentials and experience of your instructors in addition to the safety precautions that are in place. To attract new students, think about providing free trial sessions or introduction classes. Submit student success stories and progress photos of pupils who, with your help, have become more proficient riders. Potential pupils may also be drawn in by interesting content, such as instructional films.

## Services for Breeding

Should your equestrian center provide breeding services, your advertising campaign ought to emphasize the caliber of the horses and the proficiency of your breeding scheme. Display the accomplishments and pedigrees of your horses and include thorough information on the breeding procedure. To network and advertise your breeding services to prospective customers, go to equestrian events or fairs. Developing instructional materials regarding the advantages of your breeding program can also establish you as an industry leader.

**Creating a Powerful Online Identity**

Having a solid web presence is crucial for promoting your horse farm in the modern digital era. You may interact with potential customers and reach a wider audience with the aid of an efficient web approach.

### Creating An Easy-To-Use Website

Create a website that is easy to use and concisely describes your offerings to start. To create a welcoming atmosphere, include high-quality photos of your farm, the horses, and your amenities. Don't forget to include important facts like cost, availability, and contact information. You can also benefit from having a blog section where you can post advice, success stories, and ideas regarding training and horse care.

## Making use of social media
Social media sites are excellent resources for audience engagement. Make accounts on social media sites such as Facebook, Instagrammu, and Twitter so that you may communicate with fans, publish pictures and videos, and share updates about your farm. Showcase events or training sessions using live footage, and invite customers to contribute their own content by asking them about their experiences on the farm.

## Putting SEO Strategies into Practice

Use search engine optimization (SEO) techniques to improve your online presence. Look up pertinent terms associated with horse husbandry and your offerings, then include them in the text of your website. By doing this, search engines will rank your website higher, which will make it simpler for potential clients to locate you. To further show your knowledge, think about penning guides or articles that tackle often-asked issues or popular subjects among equestrians.

**In summary**

Establishing a solid internet presence, determining your target market, and advertising your services are all essential components of a comprehensive marketing strategy for horse farms.

You can successfully attract customers by customizing your marketing efforts based on your understanding of who they are and what they desire. To stand out and expand your horse farming business in today's cutthroat industry, you must make use of both traditional and digital marketing techniques.

# CHAPTER ELEVEN

## Increasing Your Expertise And Capabilities

### Sources For Additional Education

Developing your knowledge and abilities as a novice in horse farming is crucial to managing and taking good care of your horses. You may learn everything there is to know about horse husbandry by using a variety of sites.

Books: One of the best tools for in-depth education is still books. Many books address all facets of horse husbandry, from fundamental maintenance to sophisticated breeding methods. Consult your neighborhood bookshops or libraries for suggested literature, and look via internet resources like Amazon or specialty equestrian stores. Some well-known publications are "The Complete Guide to Horse Care" and "Horse Farming: The Essential Guide." These books are useful guides because they frequently contain insightful commentary from knowledgeable horse farmers.

Courses: For those new to horse husbandry, both online and in-person courses provide structured learning opportunities. Horse care, training, breeding, and farm management are all covered in the courses offered on websites such as Coursera and Udemy. Search for agricultural colleges or institutions that provide authorized programs and maybe credentials. By taking part in these courses, you can acquire vital information and useful skills that you can use right away in your farming endeavors.

Seminars and Workshops: Taking part in seminars and workshops is a great opportunity to expand your knowledge and connect with industry experts. Numerous institutions and equestrian organizations organize seminars on various facets of horse farming. Frequently, these events include knowledgeable presenters who impart knowledge and best practices. You can enhance your comprehension of the subject matter by practicing skills and approaches in real-world situations by taking part in practical workshops.

## Developing Professional And Horse Farming Networks

One of the most important ways to increase your knowledge and proficiency in horse farming is to network with other horse farmers and equestrian professionals. Developing connections within the horse farming community can result in opportunities, shared knowledge, and experiences that can improve your farming career.

Local Equestrian Clubs and Associations: Making connections with other equestrian enthusiasts and farmers can be facilitated by becoming a member of your local clubs or associations. These groups frequently provide gatherings, contests, and events where you can meet people who share your interests.

Participating in club activities promotes information sharing and camaraderie.

Online Communities and Forums: With the advent of digital platforms, horse farmers may now connect easily through online platforms. There are specialized communities on websites like Facebook and Reddit where you may post queries, exchange experiences, and pick up tips from other users. Engaging in conversations and consulting with experienced farmers can bring you perspectives not available in textbooks or educational programs.

Opportunities for Mentorship: Having a mentor in the equestrian field can greatly accelerate your learning curve.

You can get practical advice from a mentor on managing a farm, taking care of horses, and running a company. Seek seasoned farmers who are eager to impart their knowledge. Developing a relationship between a mentor and a mentee can offer you insightful knowledge and a customized education.

## Keeping Up With Best Practices And Industry Trends

Success in the quickly changing world of horse husbandry depends on keeping up with industry trends and best practices. Keeping up with the latest developments will help you raise horses with the finest care possible and enhance your farming methods.

Industry Publications and Journals: You may stay up to date on the newest trends, scientific discoveries, and industry best practices in horse husbandry by subscribing to equestrian periodicals and journals. Experts in the industry frequently write pieces for publications like Equus and Horse & Rider, offering thoughts on subjects including diet, medical treatment, and breeding developments.

Social Media and Online Resources: You may stay up to date on new techniques, items that are just getting started, and farmer success stories by following credible equestrian organizations, trainers, and vets on social media channels.

Social media sites such as YouTube and Instagram are replete with content, from live discussions on current issues in horse husbandry to informative videos. Interacting with these groups can offer guidance and motivation.

Ongoing Education: A dedication to ongoing learning is essential in the horse farming industry.

The healthcare business is always changing, with new methods, protocols, and advancements in technology appearing regularly. You can learn about cutting-edge procedures and goods by going to conferences and expos where business executives congregate. To stay educated and keep improving your abilities, think about setting aside some time each month to read articles, watch webinars, or interact with instructional information.

## conclusios

A thrilling endeavor that delivers enormous financial and personal advantages is starting a horse farming business. We have covered a number of crucial topics related to horse farming in this book, such as choosing the appropriate breed, setting up appropriate facilities, and comprehending equine care and management techniques. Now that you are a novice, you have the fundamental understanding required to make wise choices and build a successful horse farming business.

## Summary of Important Ideas

Selecting the appropriate breed for riding, breeding, or recreational purposes is a crucial first step toward being a successful horse farmer. Developing a customized approach to care and management will be made easier if you are aware of the special traits and requirements of various breeds. Building suitable facilities is also essential for your horses' health and well-being, including riding arenas, pastures, and stables.

Regular feeding, veterinarian treatment, and grooming are all part of proper horse care. To keep your horses healthy and productive, establish a regimen that includes dental care, immunizations, and health check-ups. It is possible to improve a horse's performance and general well-being by feeding them a balanced diet that meets their nutritional needs.

**Creating a Network of Support**

In the realm of horse husbandry, having a solid support system can be quite helpful. Participating in online forums, joining local equestrian clubs, and establishing connections with other horse owners can all offer insightful conversations, supportive counsel, and companionship. It can be uplifting and instructive to exchange experiences with people who are as passionate about horses as you are; it can help you overcome obstacles and recognize accomplishments.

## Ongoing Education

Keep in mind that horse husbandry is a constantly changing field as you learn more about it. Maintaining current with emerging trends, methods, and best practices requires ongoing learning. To broaden your network and increase your knowledge, think about going to seminars, workshops, and horse events. Reputable websites, books, and online resources are great places to continue your studies and gain a deeper grasp of managing and caring for horses.

## A Fulfilling Adventure

In the end, horse farming is a rewarding lifestyle that strengthens the special link between people and horses, not just a business. Your life will be enhanced by the experiences of raising healthy horses, teaching them different tasks, and feeling proud of yourself for managing a profitable farm.

As you progress on your equine agriculture expedition, always remember to use patience and flexibility. There will be difficulties, but you can overcome them and build a successful business if you have commitment, knowledge, and a passion for horses. May your love of horse farming bring you long-term happiness and prosperity.

**THE END**

www.ingramcontent.com/pod-product-compliance
Lightning Source LLC
Chambersburg PA
CBHW070956240526
45469CB00016B/1207